telling time

step-by-step

by Bridgette Sharp

Copyright © 2018 Bridgette Sharp
All rights reserved.

Published in the United States of America. No part of this publication may be reproduced, stored in a retrieval system, or transmitted, in any form or by any means: electronic, mechanical photocopying, recording or otherwise, by anyone other than the original purchaser for his or her own personal use without the prior written permission of the publisher.

The intention of this book is not to diagnose or treat any medical condition. The author is not a medical doctor. This book is meant to be used for educational purposes only. Please consult a medical professional for any medical concerns that you may have.

ISBN-13: 978-1725863224

ISBN-10: 1725863227

Table of Contents

Getting Familiar with the Clock Face	4
The Hour Hand	10
The Minute Hand	18
The Watch Face	26
Telling Time: Whole Hours	27
Telling Time: Half Hour	30
The Minutes	32
Telling Time: Five Minute Intervals	33
Telling Time Minutes	35
Telling Time Practice Pages	36
Getting Familiar	37
Write in the Missing Numbers	38
Telling Time Hours Practice	42
Telling Time Half Hour Practice	43
Telling Time 5 Minute Practice	44
Telling Time Minute Practice	45
Answer Key	47

Getting Familiar

1. Color in each of the numbers on the clock face
2. Draw a straight line from 12 to 6
3. Draw a straight line from 9 to 3

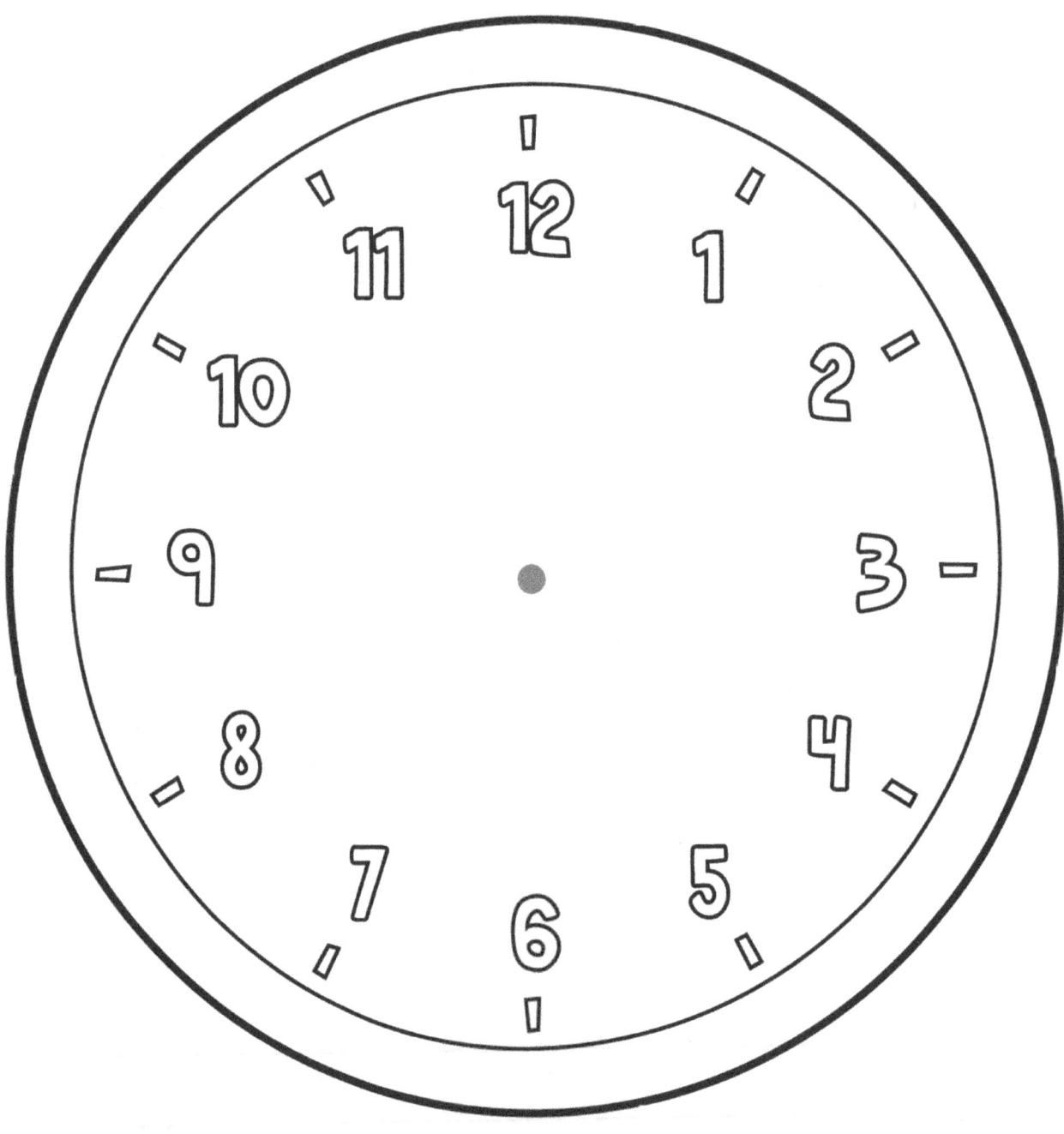

Telling Time Step-by-Step

Getting Familiar

1. Write in the missing numbers
2. Color in the remaining numbers

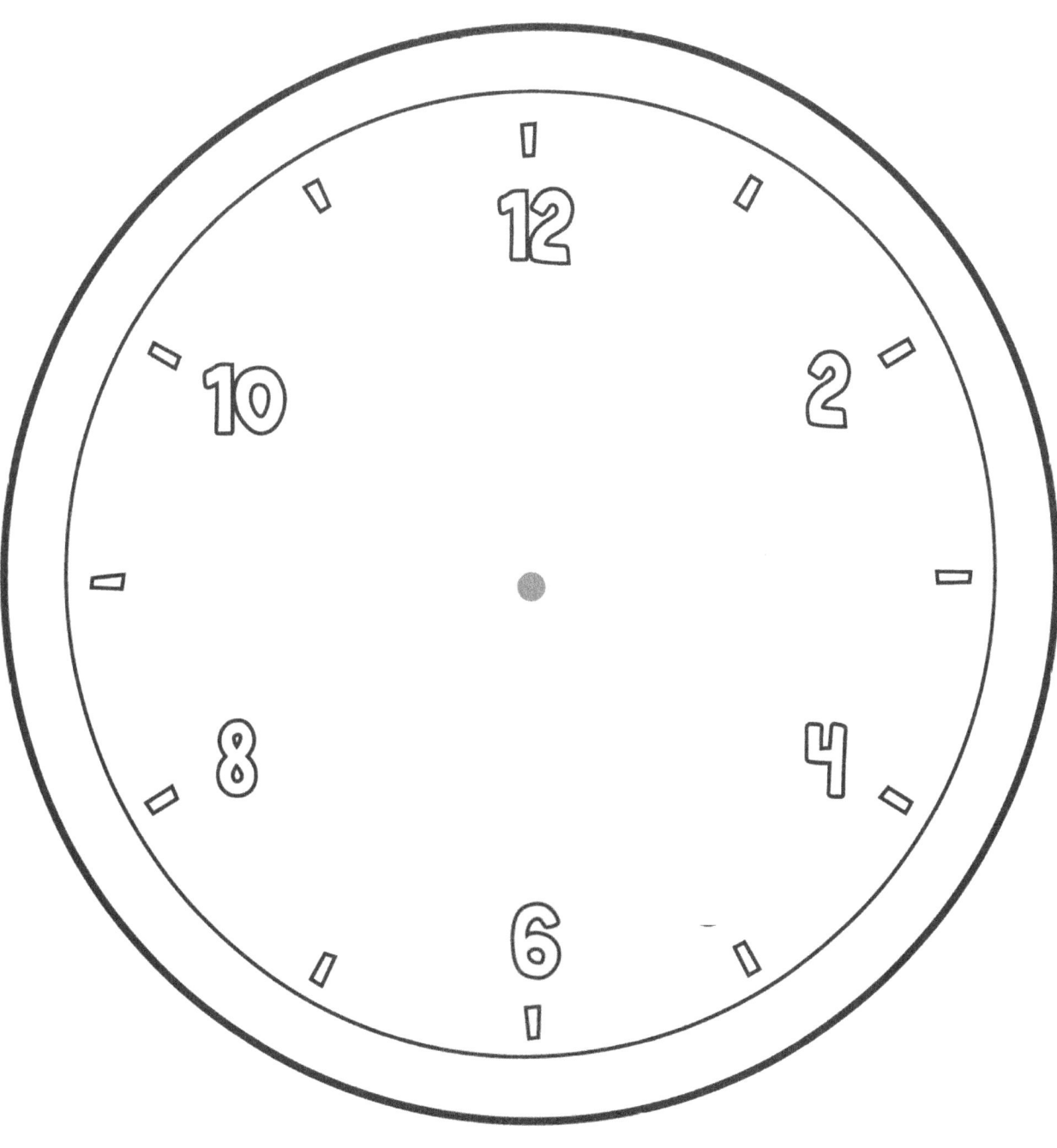

Telling Time Step-by-Step © Copyright 2018 Bridgette Sharp

Getting Familiar

1. Write in the missing numbers
2. Color in the remaining numbers

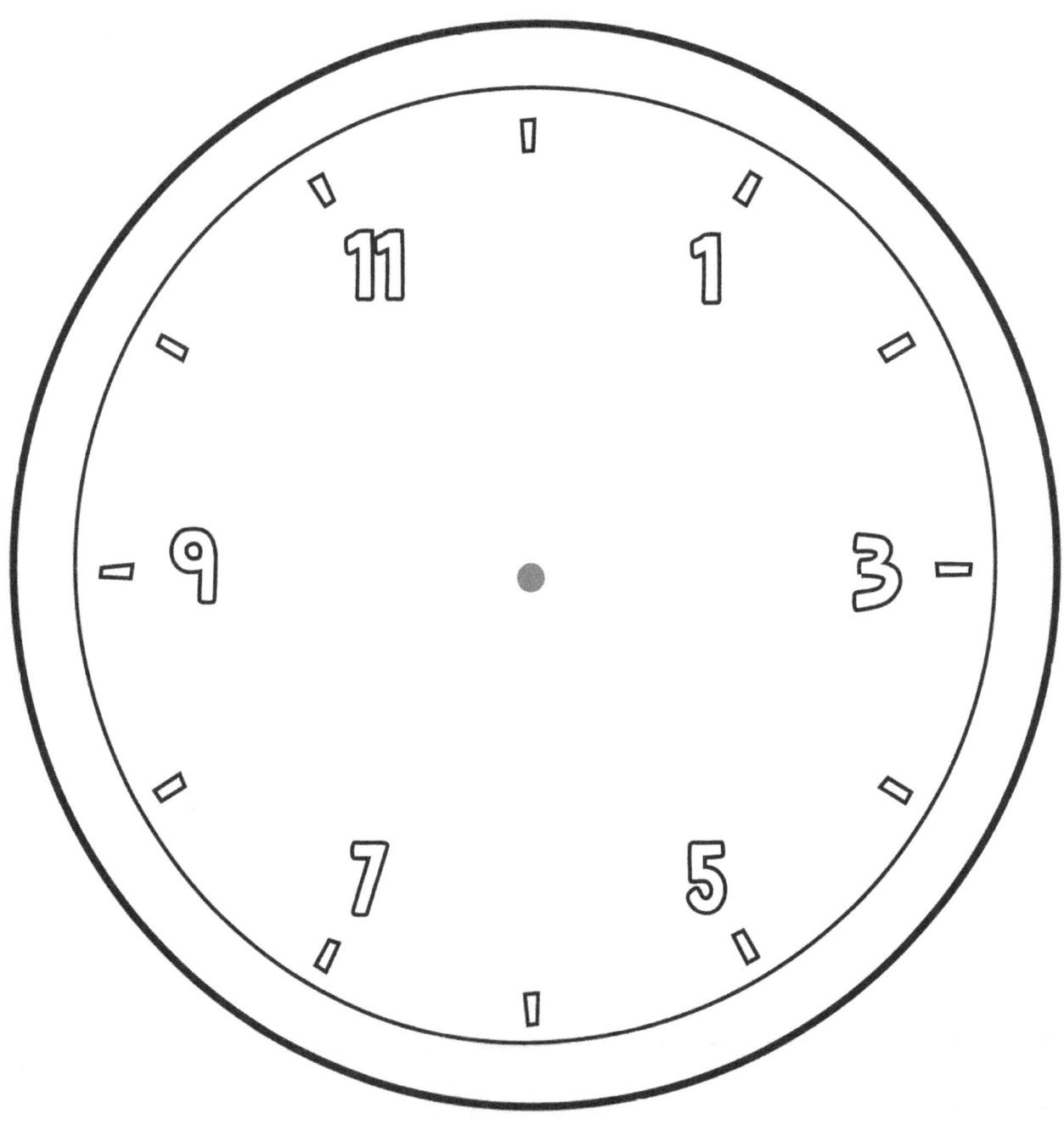

Telling Time Step-by-Step © Copyright 2018 Bridgette Sharp

Getting Familiar

1. Write in the missing numbers

Telling Time Step-by-Step © Copyright 2018 Bridgette Sharp

Getting Familiar

1. Write in the missing numbers

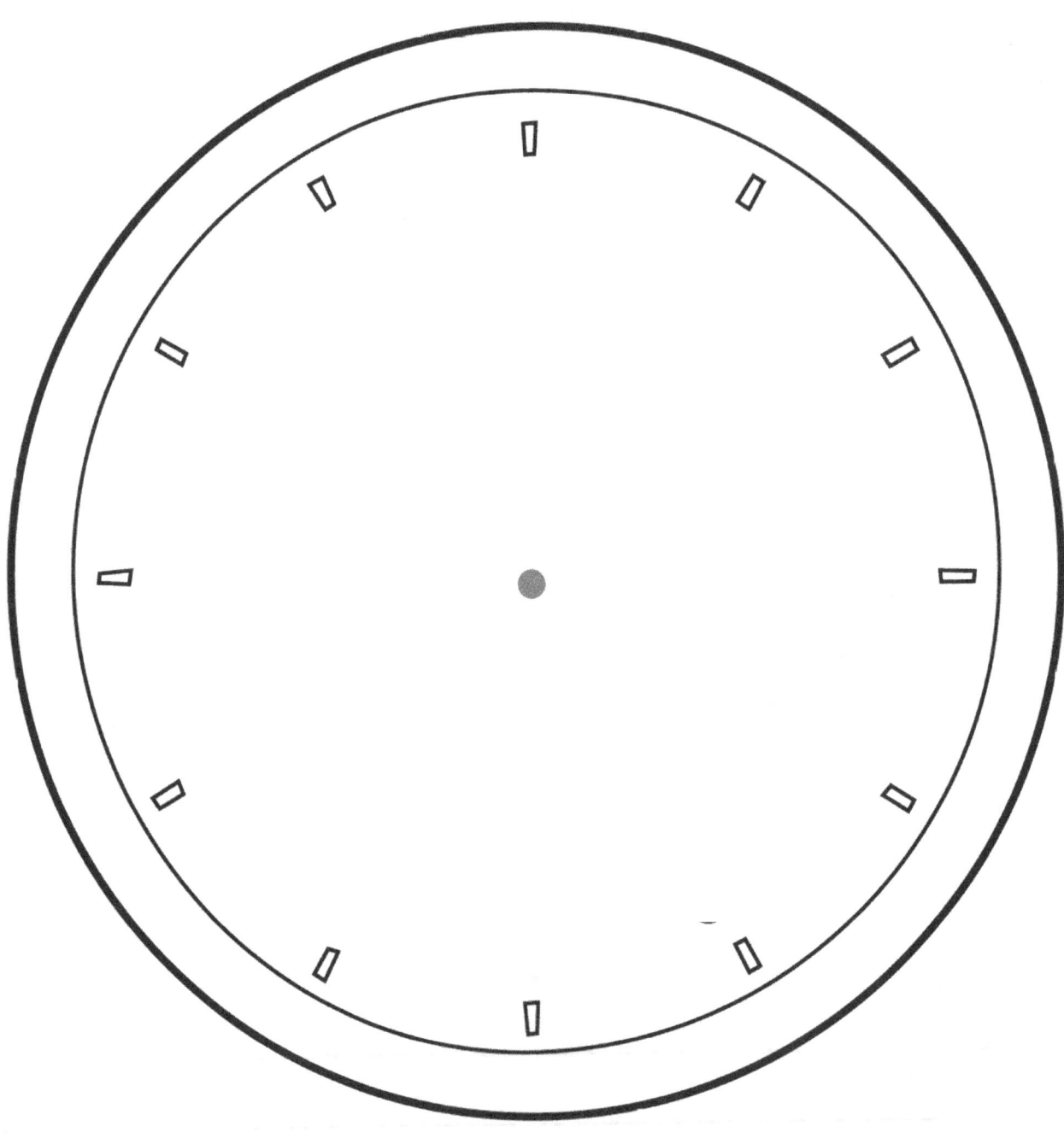

Telling Time Step-by-Step

Getting Familiar

1. Write in the missing numbers

Telling Time Step-by-Step

The Hour Hand

1. Each big number on the clock represents the hour .
2. The hour hand points to the big numbers on the clock.

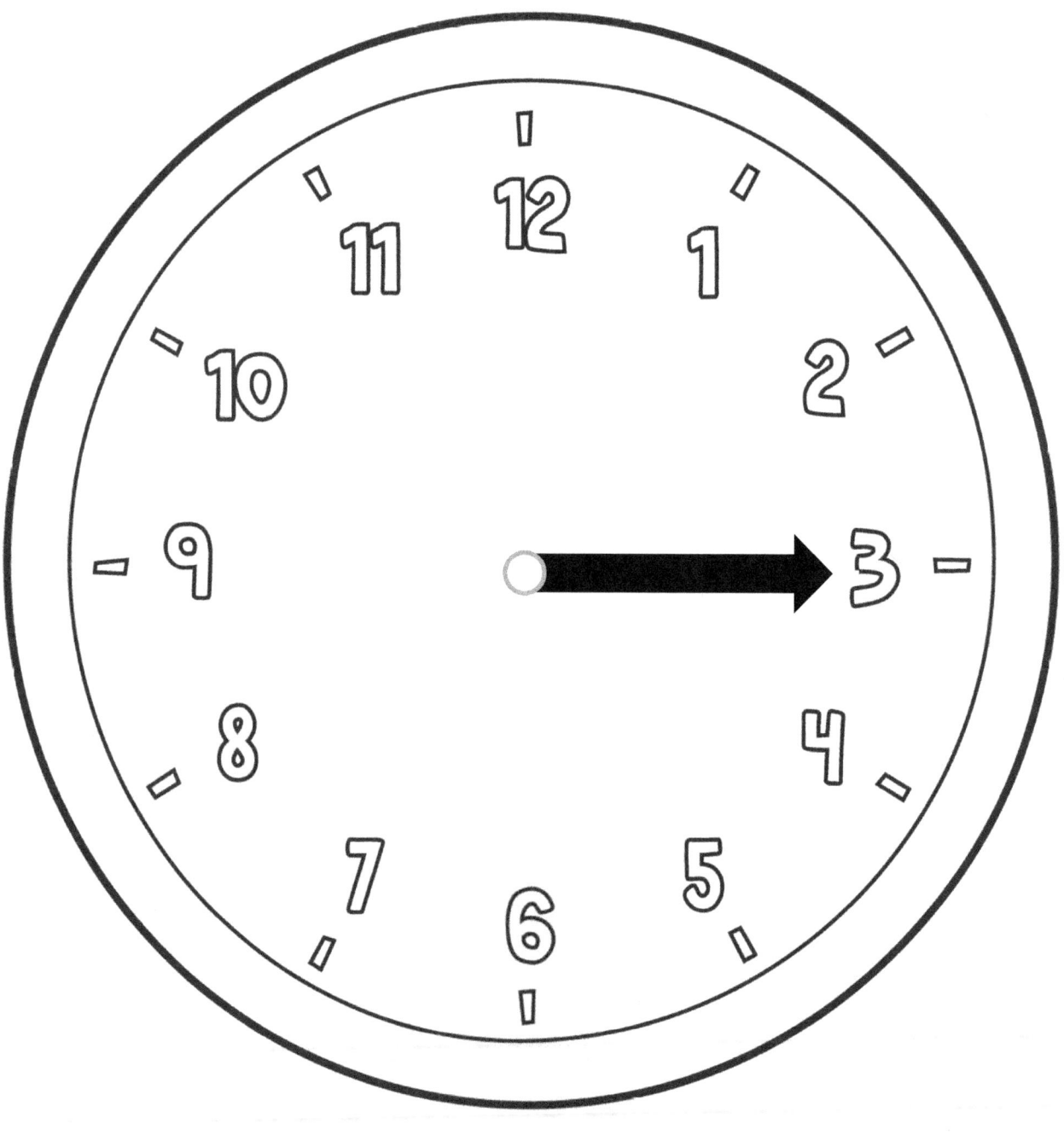

It is three o'clock; 3:00.

The Hour Hand

1. What number is the hour hand pointing to?

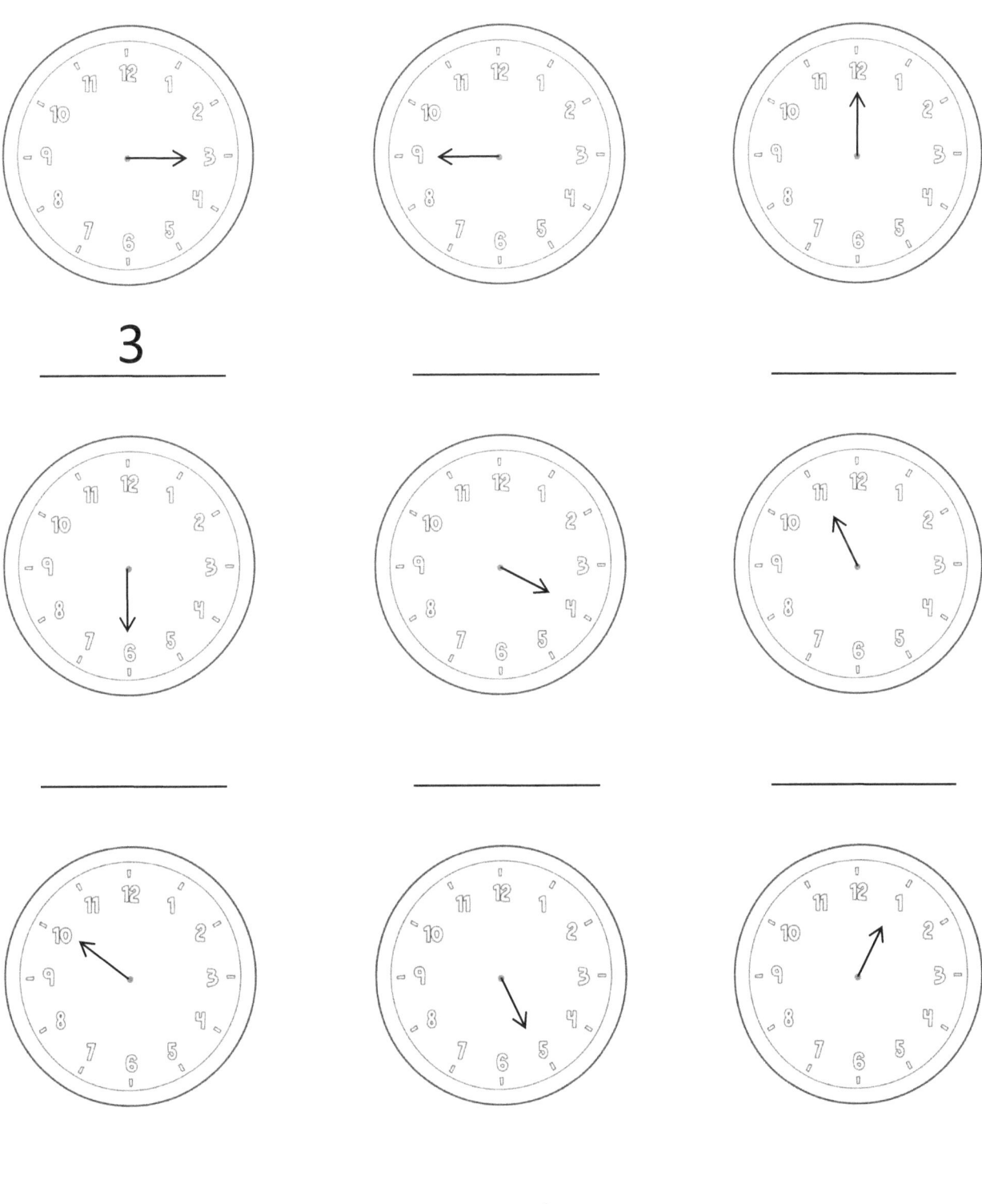

The Hour Hand

1. Draw an arrow from the middle to the number listed.

6

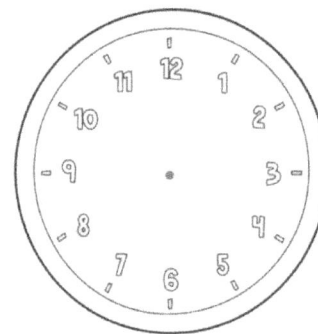
1

11

8

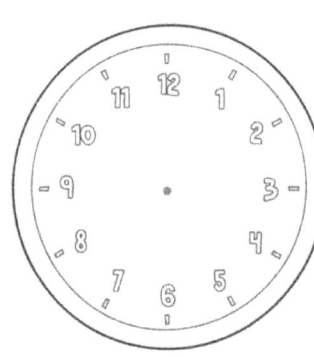
3

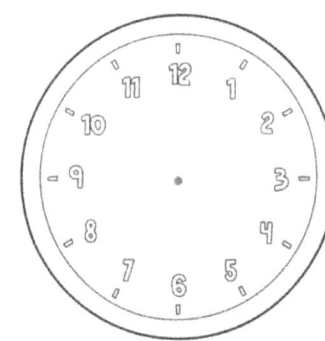
12

5

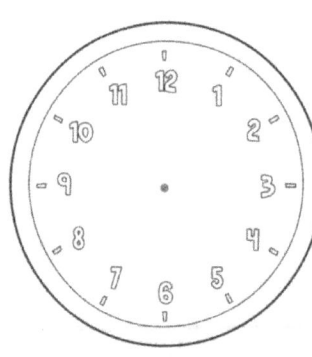
9

7

Telling Time Step-by-Step © Copyright 2018 Bridgette Sharp

The Hour Hand

1. Write the number the arrow is pointing to before the colon (dots).

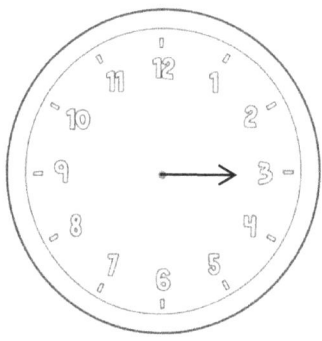
3:00

:00

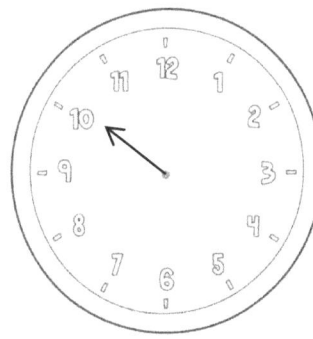
:00

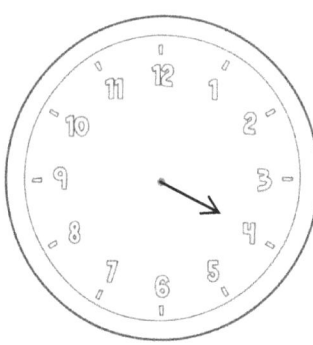
:00

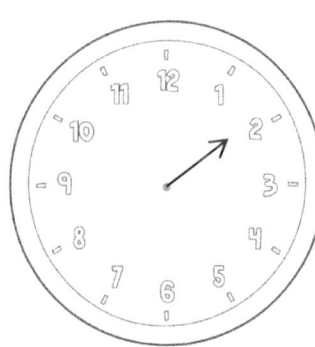
:00

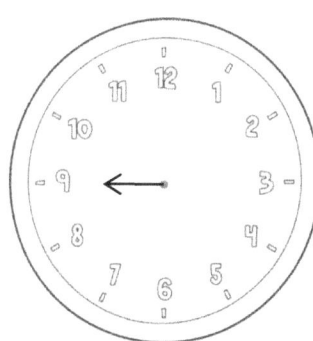
:00

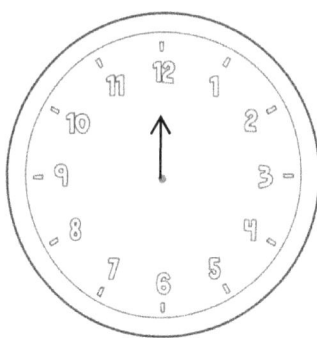
:00

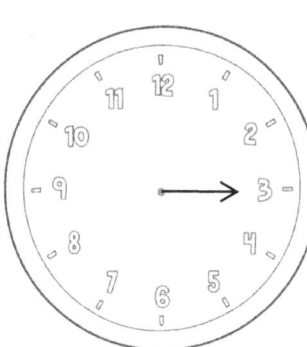
:00

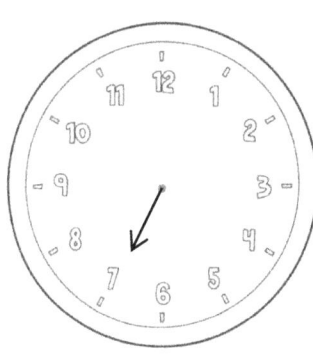
:00

Telling Time Step-by-Step © Copyright 2018 Bridgette Sharp

The Hour Hand

1. Draw an arrow from the middle to the hour listed.

6:00

12:00

3:00

9:00

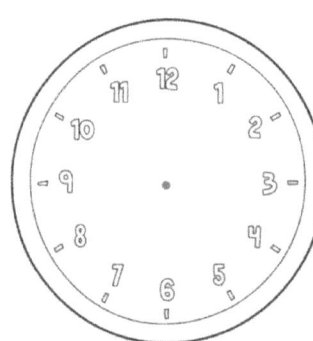
1:00

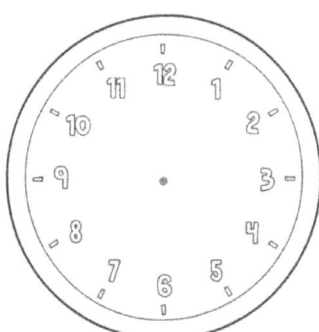
8:00

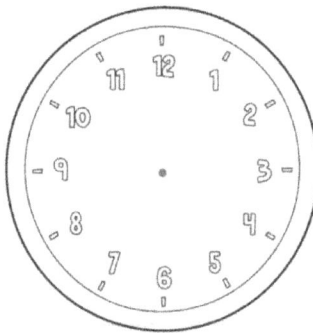
2:00

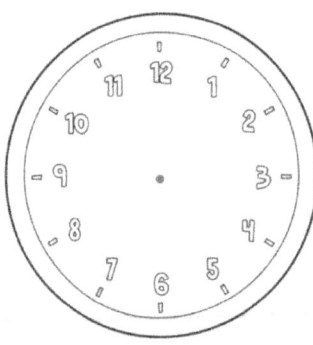
7:00

10:00

Telling Time Step-by-Step

The Hour Hand

1. Each big number on the clock represents the hour .
2. If the arrow is between two numbers, use the smaller number.

Since the arrow is between 5 and 6, we will use the number 5.   5:_ _

Telling Time Step-by-Step © Copyright 2018 Bridgette Sharp

The Hour Hand

1. Write the lower number before the colon (dots).

The Hour Hand

1. Write the lower number before the colon (dots).

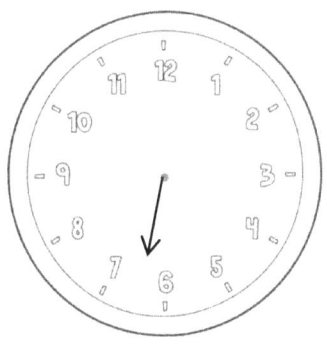
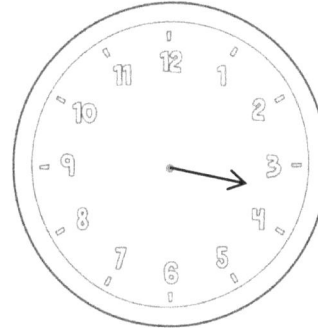

6:00   :00   :00

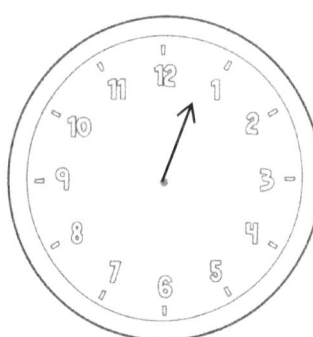
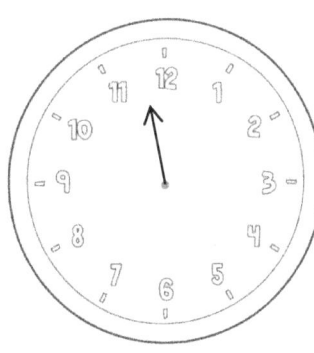
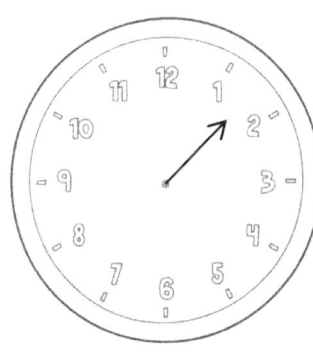

12:00   :00   :00

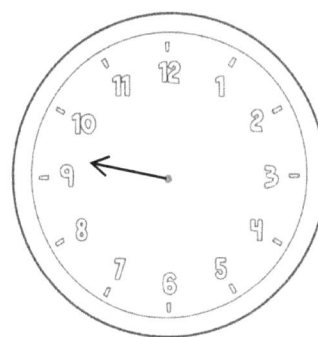
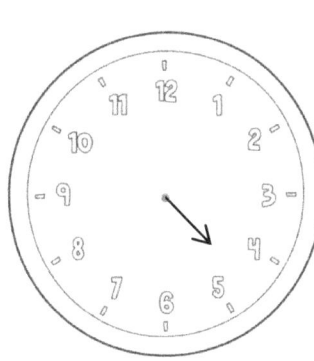
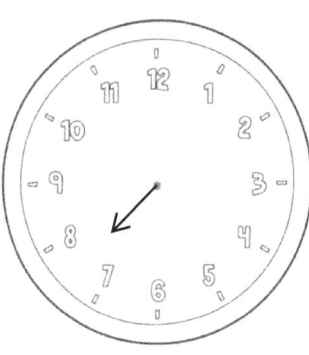

:00   :00   :00

The Minute Hand

1. Each big number on the clock represents the hour .
2. Each little number represents the minute.
3. There are 60 minutes in one hour.
4. Count the minutes

The Minute Hand

1. Each little number represents the minute.
2. There are 60 minutes in one hour.
3. Write in the minutes

Telling Time Step-by-Step © Copyright 2018 Bridgette Sharp

The Minute Hand

The minute hand points to the smaller numbers between the hour numbers.

The minute hand is pointing to 22.
It is 22 minutes past the hour. : 22

Telling Time Step-by-Step © Copyright 2018 Bridgette Sharp

The Minute Hand

1. What number is the minute hand pointing to?
2. The minute number goes after the colon (dots).

: 06

: _____

: _____

: _____

Telling Time Step-by-Step   © Copyright 2018 Bridgette Sharp

The Minute Hand

1. What number is the minute hand pointing to?
2. The minute number goes after the colon (dots).

_____ : 28 _____

_____ :0__ _____

_____ : _____

_____ : _____

Telling Time Step-by-Step   © Copyright 2018 Bridgette Sharp

The Minute Hand

1. If the minute hand is pointing through a big hour number, it is pointing to the little minute number.

_____ : 15

_____ : _____

_____ : _____

_____ : _____

The Minute Hand

1. Complete the clocks by drawing the arrow to the corresponding minute.

: 23

:30

:05

:47

Telling Time Step-by-Step © Copyright 2018 Bridgette Sharp

The Minute Hand

1. Complete the clocks by drawing the arrow to the corresponding minute.

: 00

:50

:12

:36

Telling Time Step-by-Step

The Clock Face

1. Trace the hour numbers.
2. Trace and color the hour hand (short arrow).
3. Trace and color the minute hand (long arrow).

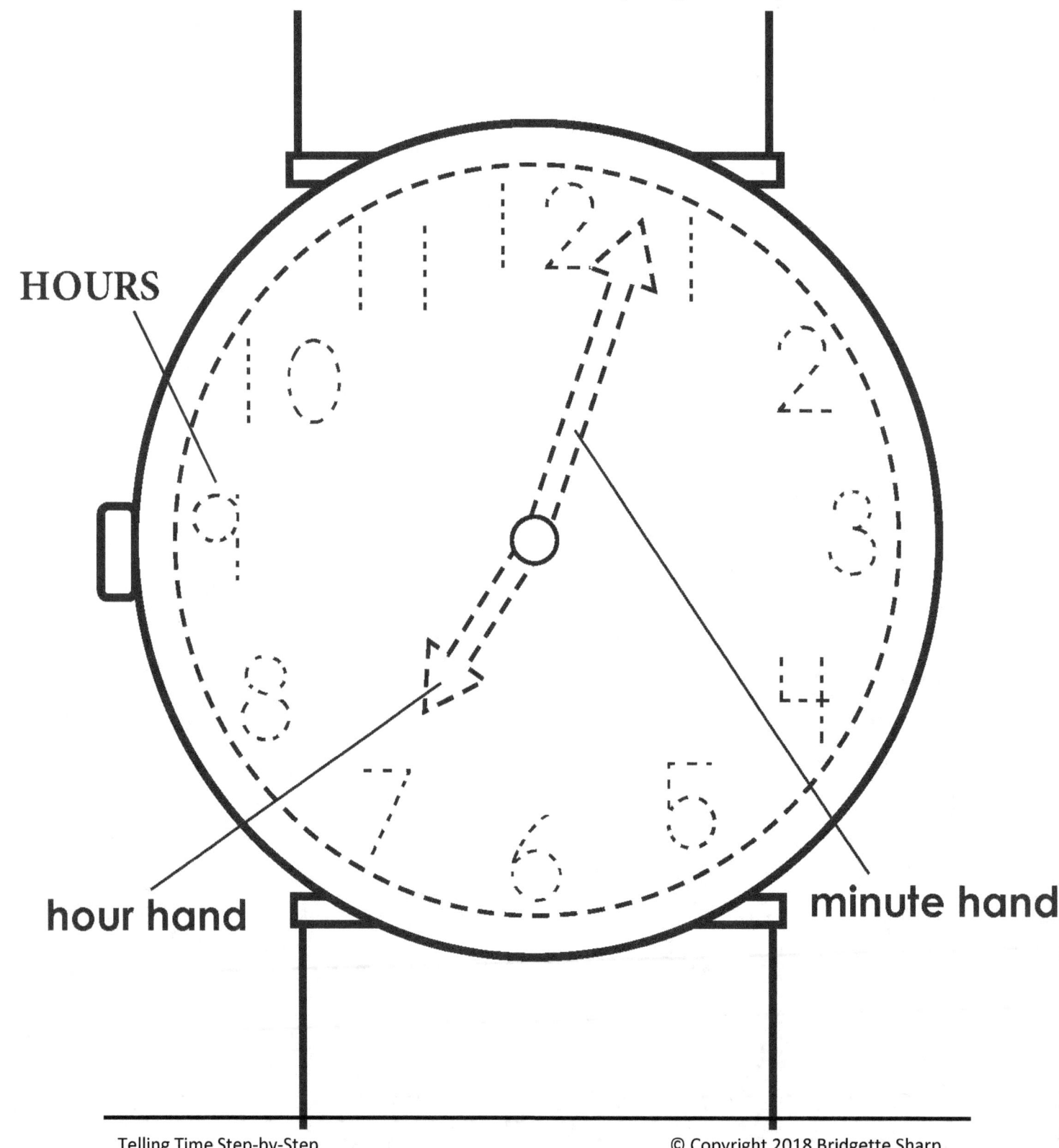

Telling Time Step-by-Step © Copyright 2018 Bridgette Sharp

Telling Time: Whole Hours

1. What number is the short hand pointing to? That is the hour. It is written before the colon.   Hour : Minutes
2. The long hand points to the minutes. It's written after the colon. If the hand points to the top, that's :00.

3:00 _____ _____

_____ _____ _____

Telling Time: Whole Hours

Write the time shown on the clock.

Telling Time: Whole Hours

Draw the hands on the clock.

Telling Time: Half Hour

1. What number is the short hand pointing to? That is the hour. It is written first.   Hour : Minutes
2. The long hand points to the minutes after the hour. It's written next. If the hand points to the bottom, that's :30.
3. Notice that at 30 minutes past the hour, the short hour hand is exactly between two hour numbers. Use the smaller hour.

5:30   _____   _____

_____   _____   _____

Telling Time Step-by-Step   © Copyright 2018 Bridgette Sharp

Telling Time: Half Hour

Draw the hands on the clock.

The Minutes
1. Can you count by 5's?   5, 10, 15, 20.....
2. Count the minutes by 5.
3. Continue drawing the lines.

Telling Time Step-by-Step   © Copyright 2018 Bridgette Sharp

Telling Time: 5 Minutes

Write the time shown on each clock. Remember to count the minutes by 5's.

_____ 3:35 _____   _____   _____

_____   _____   _____

_____   _____   _____

Telling Time Step-by-Step © Copyright 2018 Bridgette Sharp

Telling Time: 5 Minutes

Write the time shown on the clock.

Telling Time: Minutes

Write the time shown on each clock.

Telling Time Step-by-Step © Copyright 2018 Bridgette Sharp

telling time
practice pages

Getting Familiar

1. Color in each of the numbers on the clock face
2. Draw a straight line from 12 to 6
3. Draw a straight line from 9 to 3

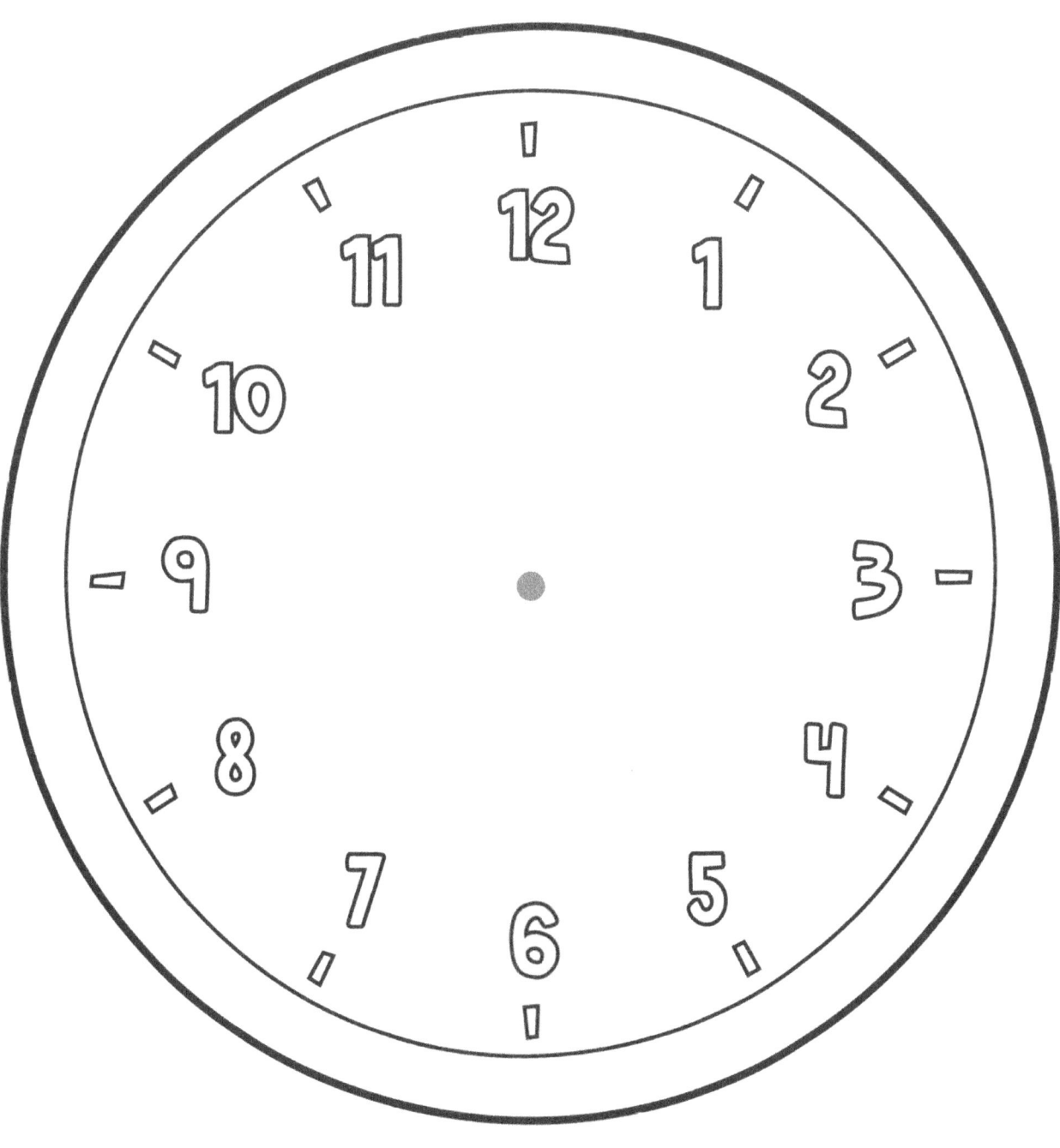

Telling Time Step-by-Step © Copyright 2018 Bridgette Sharp

Getting Familiar

1. Write in the missing numbers

Getting Familiar

1. Write in the missing numbers

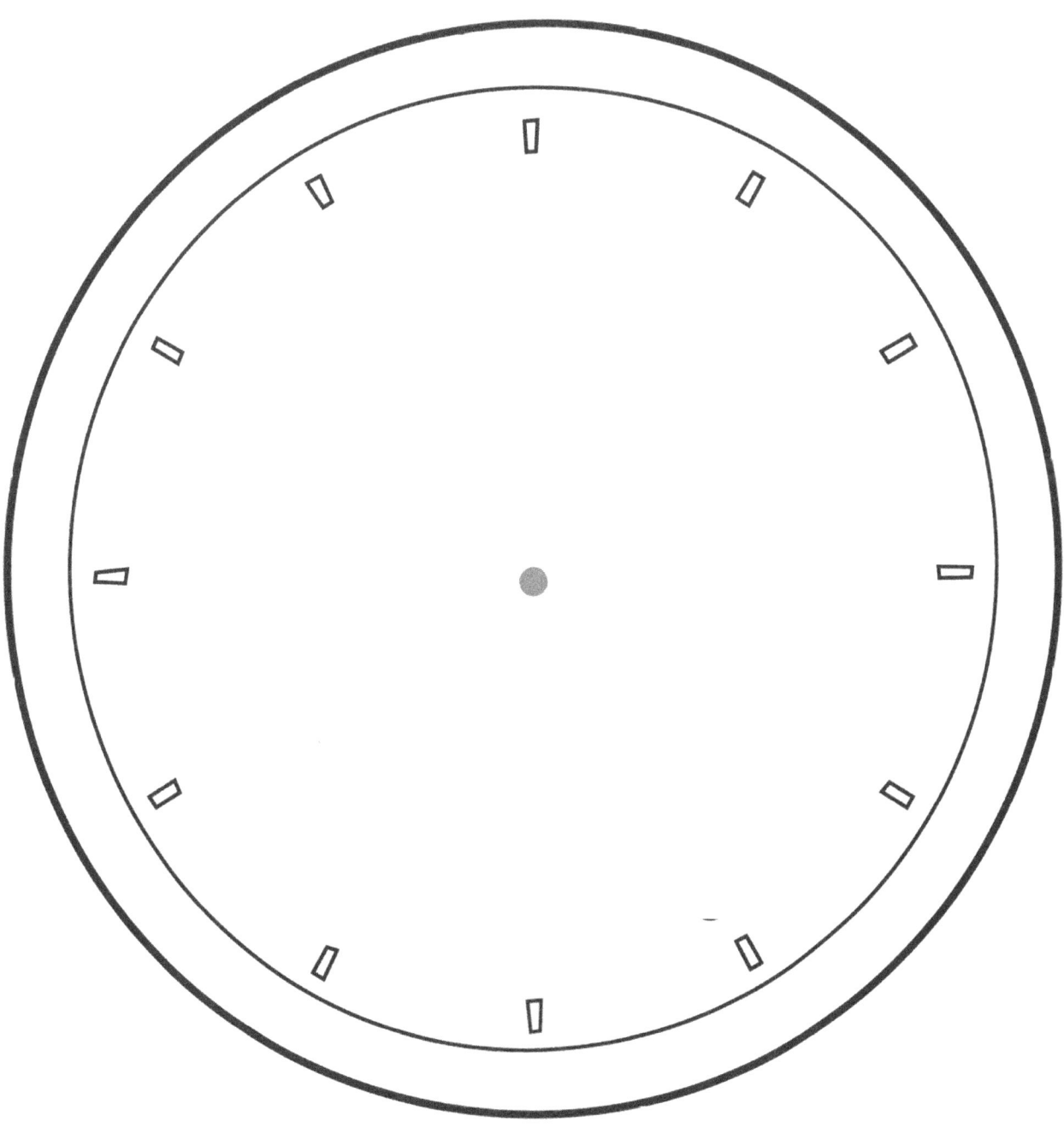

Telling Time Step-by-Step © Copyright 2018 Bridgette Sharp

Getting Familiar

1. Write in the missing numbers

Telling Time Step-by-Step© Copyright 2018 Bridgette Sharp

Getting Familiar

1. Write in the missing numbers

Telling Time Step-by-Step   © Copyright 2018 Bridgette Sharp

Write in the time.

Telling Time Step-by-Step © Copyright 2018 Bridgette Sharp

Write in the time.

Write in the time.

Write in the time.

Write in the time.

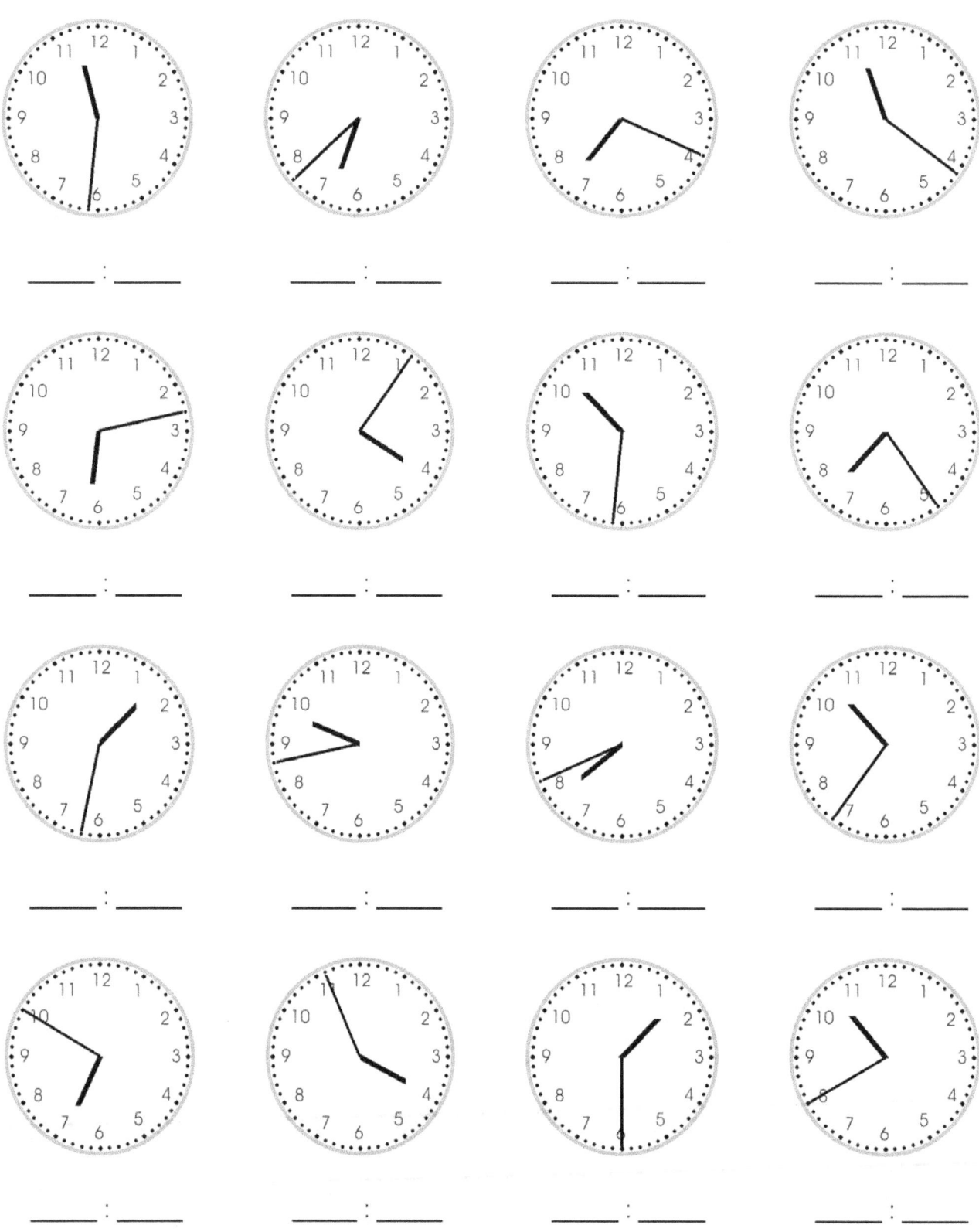

Answer Key

Page 10:	3, 9,12,6,4,11,10,5,1
Page 12:	3,6,10,4,2,9,12,3,7
Page 15:	1,7,5,3,12,10,4,8,10
Page 16:	6,3,9,12,11,1,9,4,7
Page 20:	06,38,53,13
Page 21:	28,03,43,20
Page 22:	15,30,45,10
Page 26:	3:00, 8:00, 6:00, 1:00, 12:00, 10:00
Page 27:	5, 1, 8, 10:00, 3:00, 11:00, 2:00, 4:00, 6:00
Page 29:	5:30, 4:00, 10:30, 12:30, 6:00, 8:30
Page 32:	3:35, 10:05, 1:15, 8:55, 5:25, 2:45, 12:40, 7:10, 6:20
Page 33:	5:15, 1:30, 8:50, 10:10, 3:00,11:25, 2:05, 4:35, 6:30
Page 34:	6:07, 2:33, 10:16, 4:54
Page 41:	1:00, 9:00, 7:00, 2:00, 10:00, 3:00, 12:00, 5:00, 8:00, 6:00, 4:00, 11:00
Page 42:	5:00, 1:00, 9:00, 12:30, 12:00, 11:00, 6:00, 10:00, 5:30, 8:00, 3:30, 10:30
Page 43:	8:25, 7:40, 3:40, 7:35, 9:20, 5:50 11:30, 11:45, 4;40, 3:25, 9:25, 6:40
Page 44:	4:33, 9:38, 8:22, 11:52, 1:13, 12:08, 5:28, 3:29, 9:42, 3:48, 7:26, 12:19
Page 45:	11:31, 6:38, 7:19, 11:21, 6:13, 4:06, 10:31, 7:24, 1:32, 9:43, 7:41, 10:36, 6:50, 3:55, 1:30, 10:40

Telling Time Step-by-Step © Copyright 2018 Bridgette Sharp

OTHER BOOKS BY BRIDGETTE SHARP

Decoding Sight Words Book 1 & Book 2
Read, Write & Spell Gigantic Words
30 Days of Brain Training   30 Days of Reading Drills
Brain Training Phonics: A Whole Brain Approach to Learning Phonics
Brain Training for Reversals: Letter, Number & Word Reversals
Cognitive Training Exercises
12 Weeks to Superior Memory & Mental Clarity
Brain Training Exercises to Boost Brain Power: for Improved Memory
Creative Exercises for Boosting Brain Power: Brain Balancing
Brain Training Sight Words: A Whole Brain Approach to Reading
Brain Training for Reversals: b-d-p-q
Brain Training ABC's & 123's: Kindergarten Readiness Workbook
Neuromotor Brain Training Exercises
Brain Training Sight Words: 1000 High Frequency Words Every Student Should Know
Brain Training Sight Words: 100 High Frequency Words
Hands On Phonemic Awareness Workbook
Brain Balancing Hemispheric Integration
Brain Training CVC Words
Brain Training Letter Sounds
Brain Training Numbers
Sight Word Spinners: Grades 1 - 3
Sight Word Spinners: 1000 Simple & Fun Practice for 1000 High Freq. Words
Raise Reading Scores in 5 Minutes a Day: Daily Phonics Drills
Reading Drills Kids Need to be Fast Readers
Soul Lessons Adult Coloring Book
Spirit Animals Adult Coloring Book

ONE LAST THING

If you enjoyed this book or found it useful I'd be very grateful if you'd post a short review on Amazon. Your support really does make a difference. I read all of the reviews personally and use the information to produce future publications.

www.ingramcontent.com/pod-product-compliance
Lightning Source LLC
Chambersburg PA
CBHW062343220526
45469CB00008B/2820